Why Do Bats Hang Upside-Down?

For information contact: Jennifer Kuhns, Author/Publisher
http://www.jenniferkuhns.net

ISBN: 979-8-9877485-1-0

Cover design: Jennifer Kuhns
Cover Formatting Art: Steven Kistler
Formatting: Jennifer Kuhns
Illustrations: Steven Kistler

PRINTED IN THE UNITED STATES OF AMERICA

Why Do Bats Hang Upside-Down?

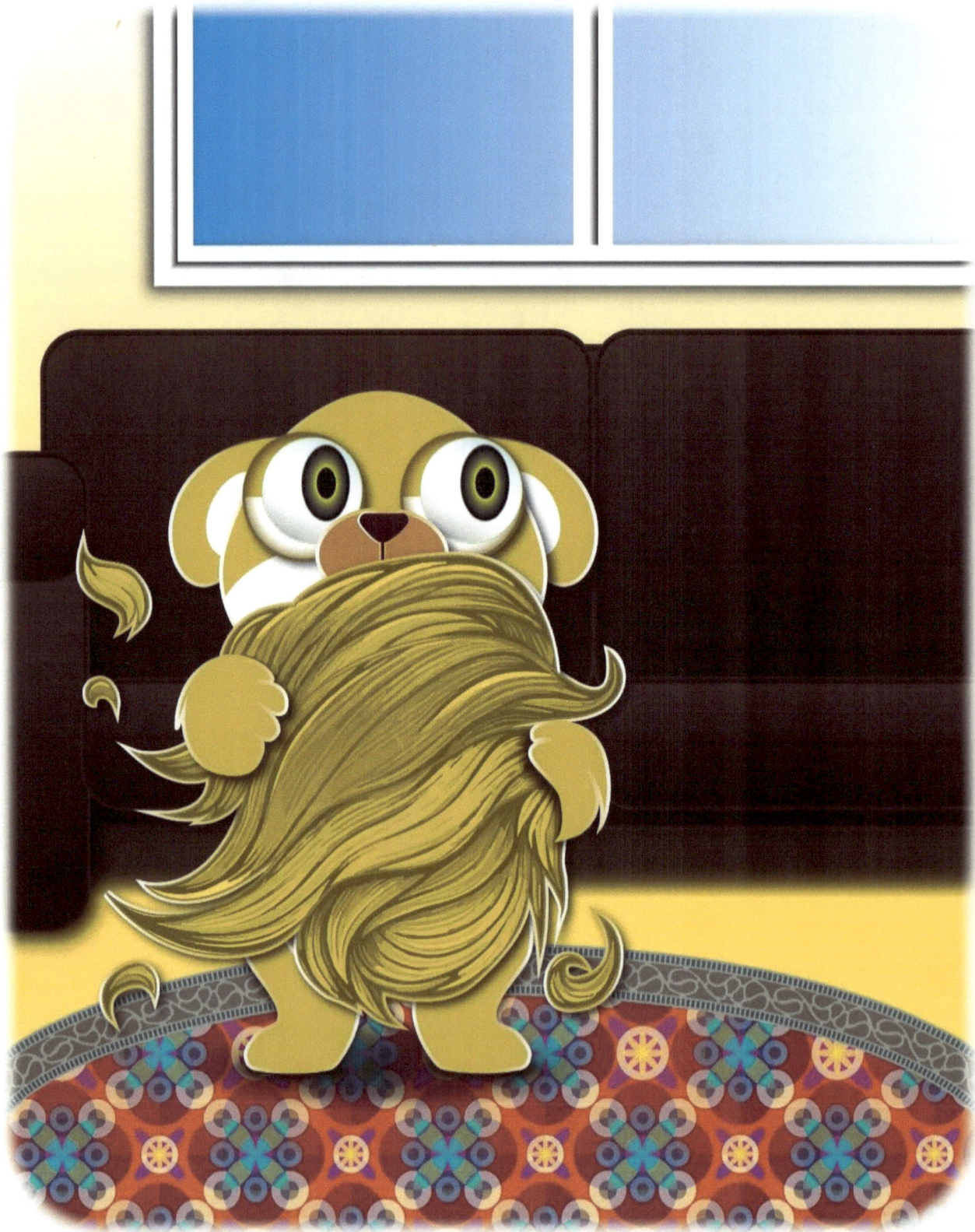

Why do cats and dogs lose their hair?

When cats and dogs lose their hair, it is called shedding. Shedding is a normal part of life for animals. Shedding helps to remove dead hair and release natural oils in the skin to keep them healthy. If dead hair does not get removed, the skin can become irritated.

Just like other animals, cats and dogs usually shed a lot twice a year because of the change of weather to either keep them cool or warm.

This usually happens in the spring and the fall; spring shedding helps pets lose their heavy winter coats, and fall shedding helps them get ready to grow winter coats.

Why do birds lose their feathers?

Actually, maintaining feathers takes a great deal of effort and energy from a bird. Every day of a bird's life, their feathers are damaged by routine wear and tear. Take a Swamp Sparrow, for example, as it finds its way through reedy marshes, it will inevitably damage its feathers by brushing them against bushes, trees and other vegetation. In addition to that, the environment birds live in, other things like bacteria, mites, lice or disease damage bird's feathers. When that happens, birds rub or pull out the damaged feathers letting new ones grow in their place.

What makes your nose itch?

There are a bazillion reasons and things that make your nose itch. Having an itchy nose is really pretty common. One reason your nose might itch is because of an allergic reaction to things like dust, pollen or animal dander (the skin flakes that come off of an animal is dander).

A couple other reason your nose might itch would be something called irritants and dryness inside the nose. Things that irritate or bother your nose and make it itchy are things like smoke, strong smells or chemical fumes. Also, dry indoor winter air dries out the inside of your nose and makes it itchy.

Even infections like colds can make your nose itch along with medications used for allergies.

Oh!!! And sticking things like Legos up your nose will make it itch, too!

What makes people stink?

You might think people stink because they sweat. But that is not totally true. Sweat, like droplets of water that come out of your skin, help people stay cool. It is when something called bacteria, things so tiny that they are invisible, that lives on a person's skin mixes with the sweat is what makes people stink.

Of course, if you don't take a bath and don't wash your clothes, sweat and bacteria mix together and build up on your skin and clothes and make you stink even more.

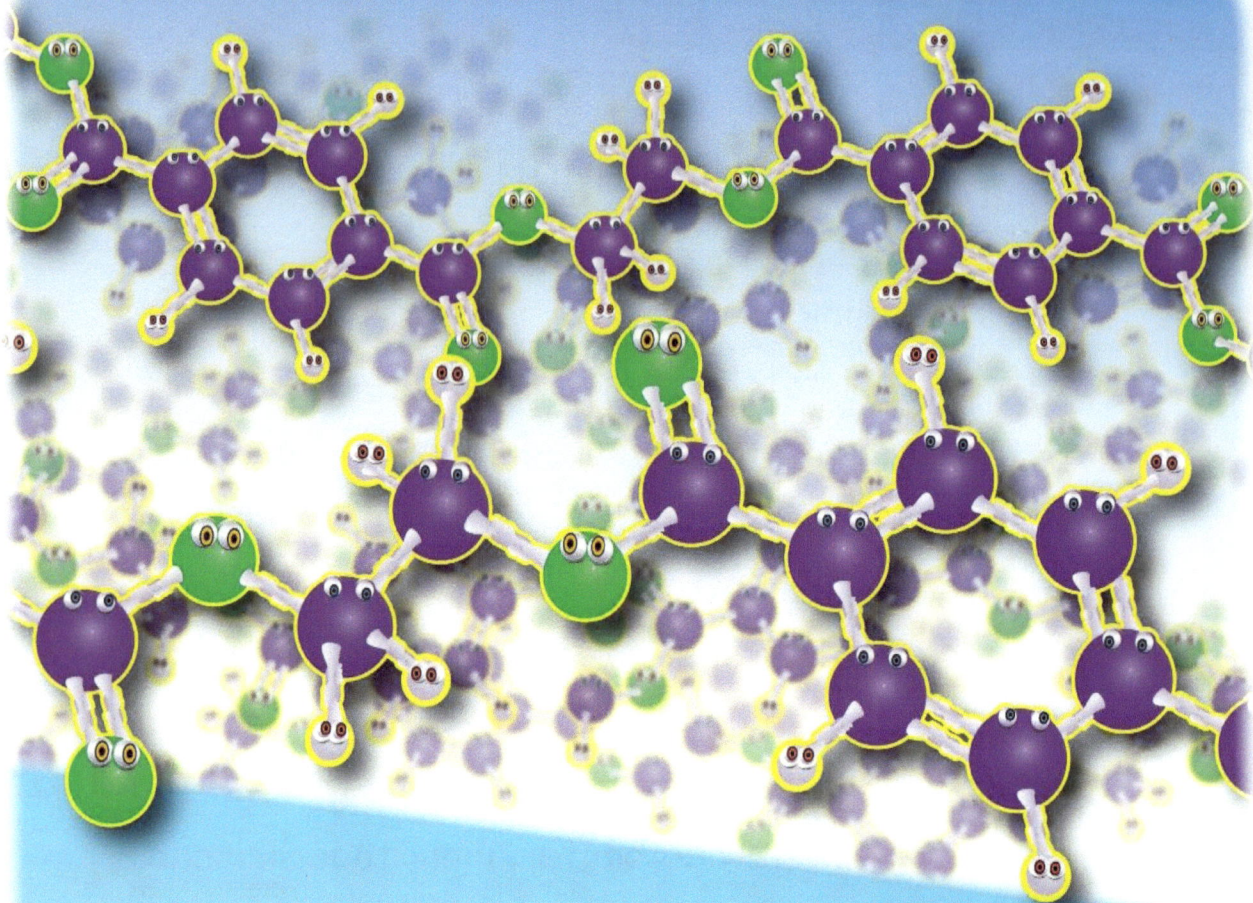

Polyester

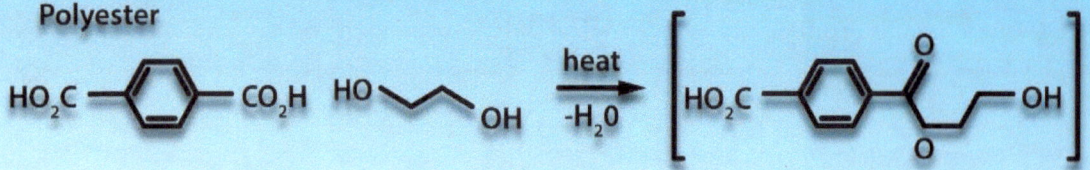

$$HO_2C \underset{}{\bigcirc} CO_2H \quad HO \underset{}{\bigcirc} OH \xrightarrow[-H_2O]{\text{heat}} \left[HO_2C \underset{}{\bigcirc} \underset{O}{\overset{O}{\bigcirc}} \underset{O}{\bigcirc} OH \right]$$

Dacron

repeat

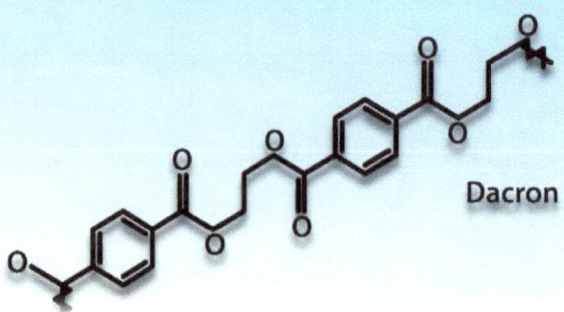

What is plastic?

Plastic is made from something called synthetic polymer. So, first let's figure out what that is...Something that is synthetic means that it didn't come from nature. It is made by chemical syntheses or change. It is the act or process of combining different things (like carbon and hydrogen for plastic). The word polymers means to combine those same thing over and over and over and over again, like the links of a chain. Plastic is a manmade material (some call it magic clay) used for all kinds of things from bottles, to dolls to even phones.

Plastic is great because it is lightweight, strong, and can last a really long time. But synthetic polymers are mostly non-biodegradable. That means they do not break down quickly or easily.

Why does hair turn white or gray?

People's hair most always turns white or gray because of the natural aging process. That is because as people get older their body makes less melanin. Melanin is a substance in your body that produces hair, eye and skin color. The more melanin your body makes, the darker your hair, eyes and skin will be.

When a person's hair might turn gray or white also depends on genetics or heredity. That means if your mom or dad's hair turned gray before they got super old, yours will probably turn gray or white before you get old, too.

Do chickens faint?

Well, this is sort of a yes and no answer. Chickens do in fact sometimes act like they faint, but it is not the same as when humans faint. Chickens can pretend to faint in order to protect themselves. Chickens can seem like they have fainted and fall over because of things like stress, being sick, sudden loud noises or being handled roughly. This is something called "tonic immobility." Tonic immobility can happen when a chicken is restrained or in a dangerous or threatening situation, like being given a bath. When a chicken faints it will be motionless, not moving, but not actually lose consciousness like a person.

How fast does ice melt?

Another tricky question . . .think about what happens when you eat and ice cream cone inside your house. As you lick it, the ice cream starts to melt because your tongue is warm. If you take the ice cream cone outside it melts faster because the sun makes it warmer outside than it is inside a house.

So, ice melts in the same way. How fast ice melts depends on how warm or hot of an environment it is in. If it is really, really hot, like on a scorching summer day, ice will melt pretty fast. But if it is winter time ice will melt much slower.

Why do leaves fall off plants and trees in the fall?

Leaves fall off trees and plants in the fall because of a combination of elements. These elements include changes in the amount of light, temperature and moisture that the trees and plants get. In the fall, the amount of daylight decreases and the temperature becomes cooler. This causes the tree to stop producing something called chlorophyll, the pigment that gives leaves their green. With less chlorophyll, the other pigments in the leaves, such as yellow and red, take over, giving the leaves their fall color. At the same time, the trees and plants stop sending energy or food to the leaves and branches which saves its energy to keep the trees and plants alive during the winter. Eventually, the leaves will die and fall off saving energy for the trees and plants to live and make it through the winter. This annual cycle is called the metabolic processes.

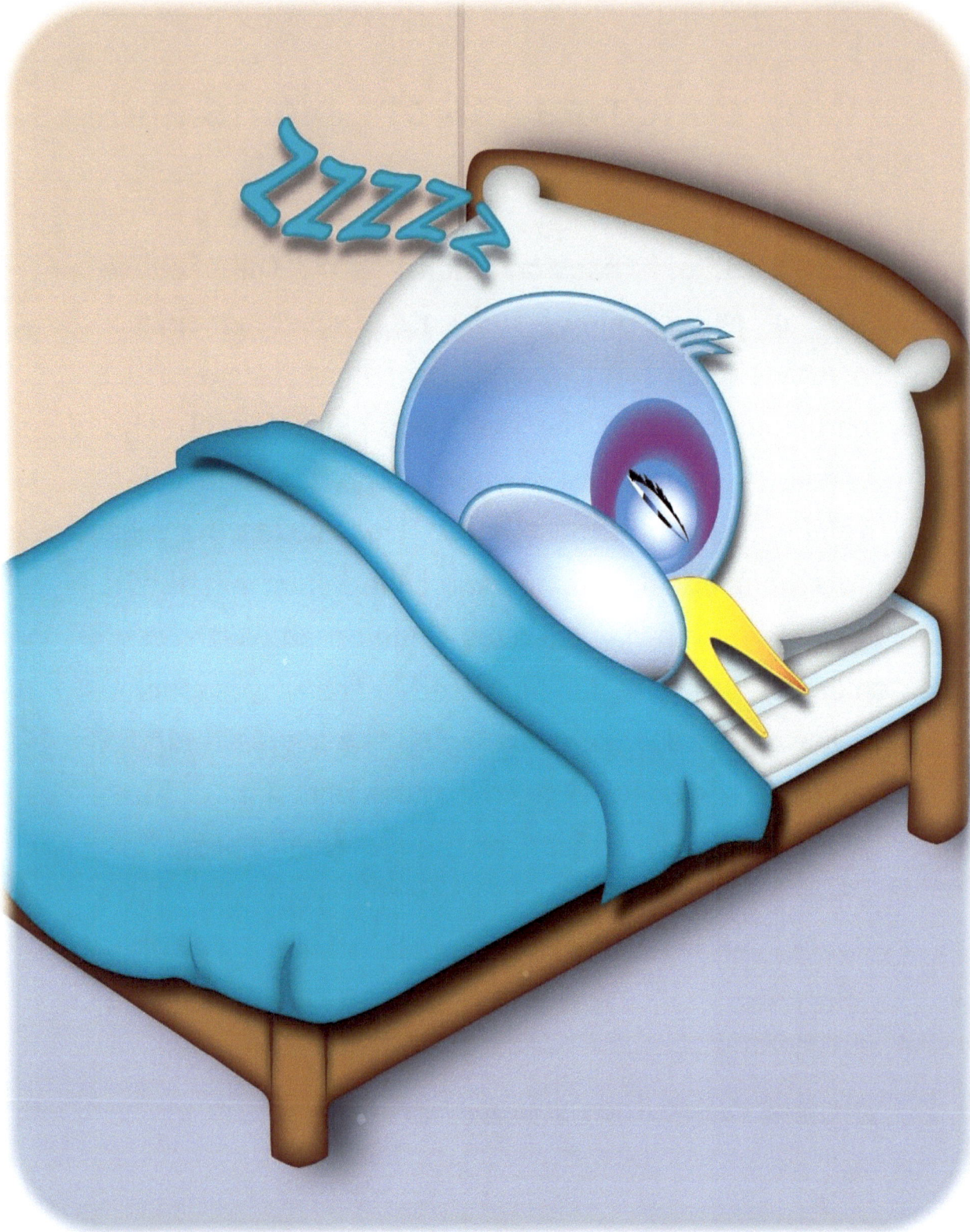

Do birds sleep?

Yes! Birds do sleep. But have you ever heard of sleeping with one eye open? Usually that is how birds sleep, with one eye open and with half of their brain awake. Birds always must be alert and aware of possible dangers, even when they are resting or asleep. This adaption or way of sleeping is called unihemispheric slow-wave sleep and keeps birds safe.

At night when it is dark and quiet and there are fewer dangers to worry about many birds will find a safe, hidden place to sleep, like tree branches or a nest where they can sleep deeper with both eyes closed and both halves of their brains resting.

Fun fact: Birds also make a blanket for themselves by fluffing up their feathers to cover their bodies to keep themselves warmer.

How does gravity work?

Gravity is a magnetic force in the middle of the earth that pulls things towards it. Imagine that there is a giant magnet in the middle of the earth and you have a metal ball in your hand. If you drop the ball the invisible magnet that is the earth's gravity pulls the metal ball to the ground. That is why things on earth don't usually float up in the air.

The same magnetic force keeps the Moon going around Earth and all of the planets in our solar system going around the sun. It is kind of what makes the whole universe stay and work together.

What is static electricity?

Everything in the world is made up of things called atoms that are so small you can't see them. Inside the atoms there are even smaller things called electrons. Sometimes atoms like to hold onto their electrons and sometimes they are willing to share.
   When atoms share their electrons, the electrons move from one place to another. For example, when you rub a balloon on your hair, some of your hair's electrons move to the balloon. Now the balloon has extra electrons and your hair has less. These are called positive and negative charges and are opposite each other. Opposite charges are attracted to each other. So, when you hold a balloon that you have rubbed on your hair next to your head your hair stands up because all the electrons want to be close together.

What makes thunder and lightning?

In rain clouds there are little drops of rain and maybe even snow that bounce around and bump into each other. When that happens, they create an electrical charge called lightening (same as static electricity). Most lightning just hangs out in the cloud and we usually don't even notice it, but every so often some leaks out and shoots down to the ground. The bright thing you see, the lightening, is only as wide as your finger, and hotter than the sun. Thunder is a sound and sound moves slower than the energy in the cloud that makes lightening. As the lightning shoots through the air super-fast, like sixty miles a second, it heats up that air. When air gets really hot, really fast it explodes. That is why you hear thunder after you see the lightening.

When and why were neckties invented?

There doesn't seem to be a clear time and reason for the invention of the necktie. It is believed that something resembling a necktie was used as far back as the second century by Chinese soldiers as a functional piece of clothing to keep their necks warm. But, since the necktie only appears on a few soldiers, it is thought by historians that the cloth was possibly an honorary badge for bravery. Also, a form of neckties was seen on Roman Croatian soldiers in the 17th century as part of their uniforms. In this case, they were similarly worn to protect their necks.

Around the same time, in the 17th century, neckties started to become popular as a fashion statement among the European upper class. Initially it was called a "cravat" and represented the idea of elegance, power and wealth. Today they are mostly a fashion statement.

Why do some animals like elephants have tusks?

First of all, only mammal have tusks. Research shows that the first mammals that had tusks were called dicynodonts that were around even before dinosaurs, and the tusks were first just big teeth.

Those big teeth grew into tusks because of something called Evolution, the idea of Natural Selection, Genetics and something called Adaptation. All of those things together caused change and development of tusks over a really, really long time; like thousands and millions of years.

Tusks are very useful to mammals like elephants for finding food and defending themselves against predators. Today's elephants have ancestors that had tusks that helped them survive. Have you ever heard the saying "that only the strong survive?"

Do animals talk to each other like humans?

Believe it or not, the answer is yes. When animals talk to each other, it is far more human-like than you might expect. According to a completely new study, many species take turns in their conversations, just like we do. Animals, like humans, seem to know when to speak and when to listen. This is called turn-taking behavior. At this point in time, no one really knows what animals say to each other.

Just as an FYI: We actually know very little about the origin of the human language.

Why do bats hang upside-down?

First, you need to know that bats are not birds. They are mammals; the only flying mammals. Birds have hollow bones. Bats do not. Birds have powerful wings. Bats do not. The difference between bat flight and bird flight is weight. So, the problem with flying for bats is that they are too heavy to take off from a standing or motionless position. This is called "the ratio of weight to lift capacity of the wing." Phew . . .that's a mouthful. To make up for the bat's inability to jump up and fly like a bird, Mother Nature (evolution) decided to hang them upside-down so they could drop from a branch and just

start flapping. Once in the air, bats can literally fly circles around most birds in flight. The problem is in first getting off the ground.

*Wondering how bats poop and pee while hanging upside-down? Pooping is no problem. They just do. Pee they have "to hold" until they are flying.

How are stars made?

Stars are not actually made; they sort of make themselves. Remember gravity? Stars become stars because of the powerful force of gravity. In galaxies there are very large and fluffy clouds of gas and dust called nebulae. Gravity clumps together some of the gases and dust inside these fluffy clouds - like raisins in a cake. When one of these clumps start to get tightly squished together, we say its density goes up. Density means how tightly something is compacted, or squished together.

These clumps of gas and dust also get hotter and hotter in the middle. When the gas and dust clumps reach a certain temperature (millions of degrees), there is a reaction inside the clumps: hydrogen atoms come together to form something called helium. When that happens something else called nuclear fusion occurs. That means a giant amount of energy is released . . .an explosion that blows apart the clumps into little bitty clumps or stars.

How high does a thermometer go?

That is a hard question because there are different kinds of thermometers. Each type of thermometer has its own range of temperatures.

   A <u>Mercury</u> thermometer used to be the most traditional kind of thermometer. It can be used to measure the temperature of almost anything. It can measure temperatures up to 660 degrees Fahrenheit.

   An <u>Alcohol</u> thermometer can measure temperatures from 100-110 degrees Fahrenheit.

   <u>Digital</u> thermometers come in a variety of types. Each has a different temperature range. Most are used for temperature of humans and have a range of 2 degrees Fahrenheit to 200 degrees Fahrenheit.

   <u>Infrared</u> thermometers measure temperatures from 50 degrees Fahrenheit to way over 1,000 degrees Fahrenheit . . .and it doesn't even have to touch anything. There are also ones that measure temperatures up into the thousands of degrees.

Humans

Cats

Dogs

Do animals see color?

Yes. Some animals see color. Although, animals see colors differently than humans. Some animals can actually see colors like humans can, some see better than humans and others don't see color as well as humans or at all.

Dogs, cats, rats and rabbits, for example, see shades of blue, yellow and gray. They can't see the color green or red like humans can.

Birds, on the other hand, can see more colors than humans. They can see something called ultraviolet colors that humans can't.

Bees and butterflies can also see a wide range of colors. They even use the color in their wings to communicate with each other.

It's like animals see their own version of the world around them. Each look at the world differently.

More children's books by Jennifer Kuhns

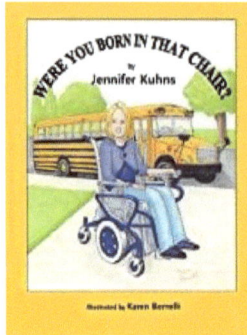

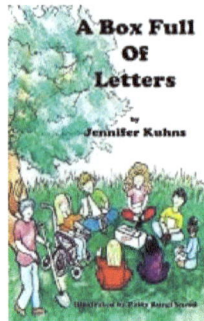

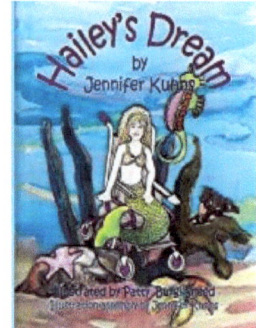

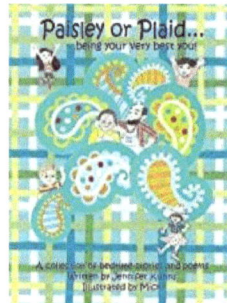

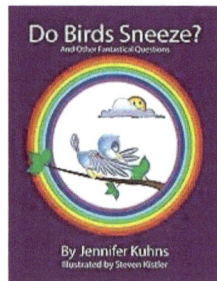

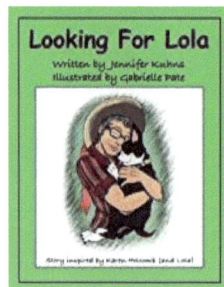

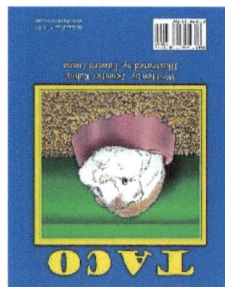

WERE YOU BORN IN THAT CHAIR?
by
Jennifer Kuhns
Illustrated by Karen Bernelli

A Box Full
Of
Letters
by
Jennifer Kuhns
Illustrated by Patty Baird Snead

Hailey's Dream
by
Jennifer Kuhns
Illustrated by Patty Baird Snead

Paisley or Plaid...
being your very best you!
A collection of bedtime stories and poems
Written by Jennifer Kuhns
Illustrated by Micki

Do Birds Sneeze?
And Other Fantastical Questions
By Jennifer Kuhns
Illustrated by Steven Kistler

Looking For Lola
Written by Jennifer Kuhns
Illustrated by Gabrielle Pate
Story inspired by Karen Maxwell (and Lola)

TACO
Written by Jennifer Kuhns
Illustrated by Jennifer Kuhns

A two-sided flipbook

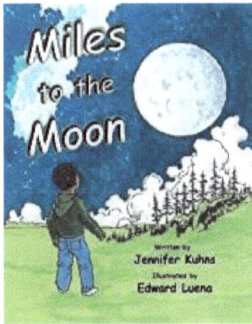

Miles
to the
Moon

Written by
Jennifer Kuhns
Illustrated by
Edward Luena

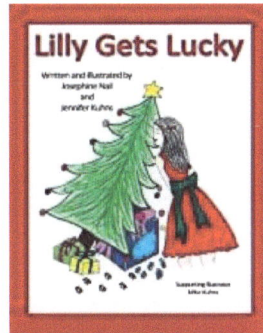

Lilly Gets Lucky

Written and Illustrated by
Josephine Nail
and
Jennifer Kuhns

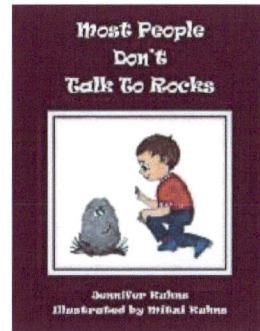

Most People
Don't
Talk To Rocks

Jennifer Kuhns
Illustrated by Mikal Kuhns

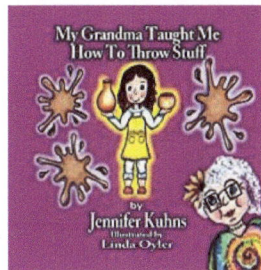

My Grandma Taught Me
How To Throw Stuff

by
Jennifer Kuhns
Illustrated by
Linda Oyler

For the adult reader

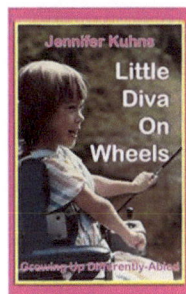

Jennifer Kuhns
Little
Diva
On
Wheels

Growing Up Differently-Abled

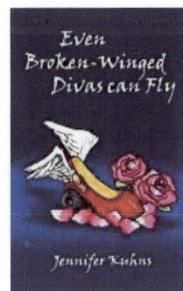

Even
Broken-Winged
Divas can Fly

Jennifer Kuhns

Do you have a fantastical question? Write it here and look for the answer.

_____

_____

_____

_____

_____

_____

_____

_____